FISH

Rebecca Woodbury, Ph.D., M.Ed.

Gravitas Publications Inc.

FISH

Illustrations: Janet Moneymaker

Fish
ISBN 978-1-950415-64-9

Published by Gravitas Publications Inc.
Imprint: Real Science-4-Kids
www.gravitaspublications.com
www.realscience4kids.com

RS4K

Photo credits: Cover & Title Pg: vlad61_61, AdobeStock; Above, Zac Wolf, CC BY SA 2.5; P.3. NOAA/NMFS/ Pacific Islands Fisheries Science Center Blog; P.5. 1) Nick Hobgood, CC BY SA 3.0; 2) Photo by David Clode on Unsplash; 3) Tiit Hunt, CC BY SA 3.0; P.11. Photo by Gerald Schömbs on Unsplash; P.15. Ali Abdul Rahman on Unsplash; P.17. Top, USFWS-Pacific Region, CC BY SA 2.0; Bottom, NOAA Great Lakes Environmental Research Laboratory, CC BY SA 2.0; P.19. NOAA Great Lakes Environmental Research Laboratory, CC BY SA 2.0

Fish are animals that swim in water. They are found in lakes, rivers, streams, and oceans.

It would be fun to swim like that.

There are three types of fish.

Bony fish

Cartilage fish

Fish without jaws

Bony Fish

1

Cartilage Fish

2

Fish Without Jaws

3

Bass are a type of **bony fish.**

Bony fish are the most common type of fish. They are named for their **skeletons**, which are made of bones.

I have a skeleton made of bones.

Bass

Skeleton made of bones

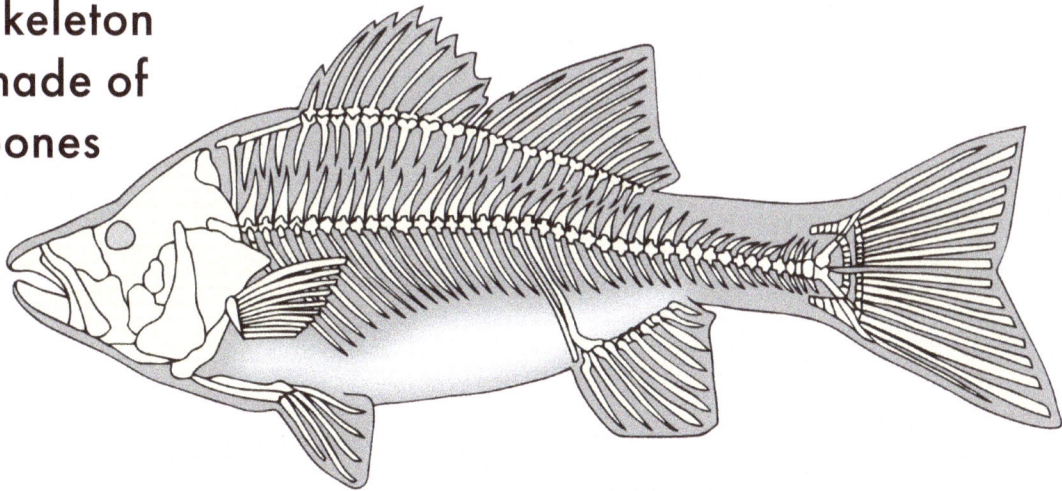

Bony fish can float because they have a **swim bladder.** They fill the swim bladder with air when they want to go higher in the water. They let some of the air out of the swim bladder when they want to sink deeper into the water.

Swim bladder

Sharks are a type of **cartilage fish.**

Sharks have a **skeleton** made of **cartilage.** Cartilage is strong and can bend. It is light in weight.

Shark

Skeleton made
of cartilage

Sharks don't have a swim bladder. Cartilage is lighter than regular bones, which makes it easy for sharks to float and swim without a swim bladder.

A **lamprey** is a type
of **jawless fish**.

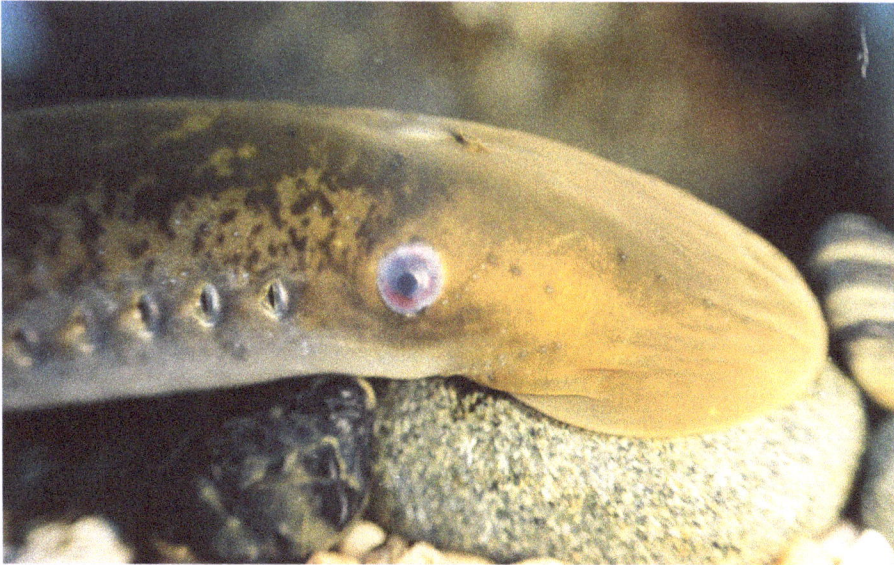

Lampreys have long thin bodies. They don't have a jaw. They eat their food by using a round mouth lined with lots of teeth.

I'm glad I don't have to brush all those teeth!

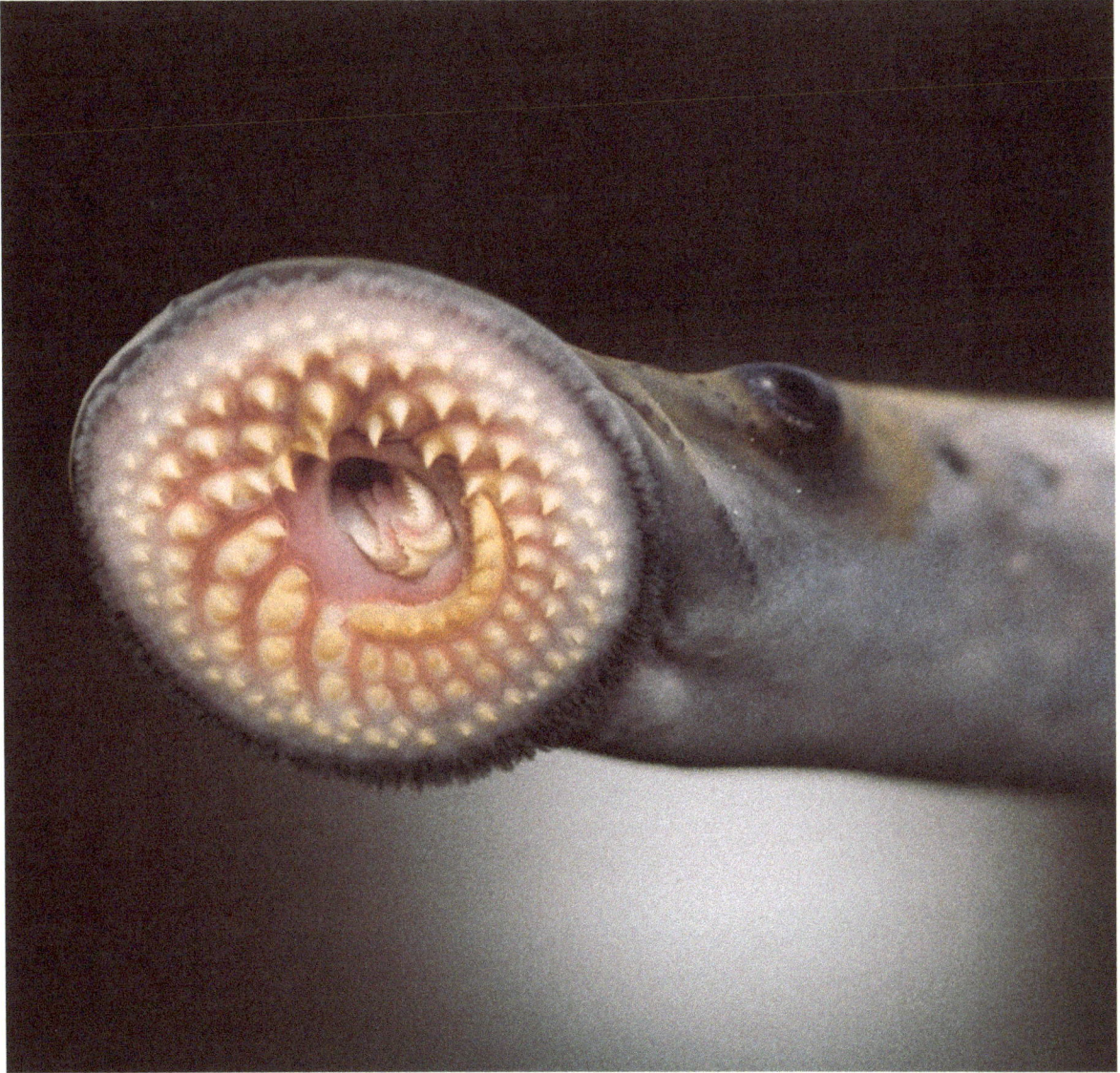

Fish are an important food for people and some animals.

Most of the fish we catch and eat are bony fish, including salmon, tuna, and trout.

Just give me a cheese sandwich, please.

How to say science words

bass   (BAASS)

bony   (BOH-nee)

cartilage   (KAHR-tuh-lihj)

jawless   (JAW-luhs)

lamprey   (LAAM-pree)

salmon   (SAA-muhn)

science   (SIY-uhns)

skeleton   (SKEH-luh-tuhn)

swim bladder   (SWIM BLAA-duhr)

trout   (TROWT)

tuna   (TOO-nuh)